NEWBERRY

The Life and Times of a Maine Clam

VINCENT G. DETHIER

illustrations by Marie Litterer

Down East BOOKS
CAMDEN, MAINE

Second printing

ISBN 0-89272-085-9

Library of Congress Catalog Card Number 81-66267

Design: Lurelle Cheverie

Printed in the United States of America

CONTENTS

NEWBERRY

1

The New Muffler

Once there was a clam named Newberry who lived in a mudflat off the coast of Maine. Life on the whole was pleasant except that the mudflat was rather chilly sometimes and rather damp most times. This suited Newberry until one particular morning in June when he woke up with a pain in his neck. As the days went on the pain got worse and worse. Not only did it hurt, it almost cost him his life on one occasion. It happened this way.

Newberry's mudflat, being along the coast of Maine, was a rocky one. This was a good thing for him and his numerous cousins because people who dug for clams preferred nice oozy mudflats without rocks. Rocks made for difficult digging. So Newberry lived a relatively safe existence. Being a cautious soul, he lived under the edge of a particularly large granite boulder. When the tide was out, he liked to push his neck out as far as it would stretch on the mud. He could stretch his neck a tremendous long way. He thought of the boulder as his boulder although it did not any more belong to him than it did to the barnacles who lived on top of it.

Newberry enjoyed stretching his neck way out on the mud because while the tide was out the sun warmed the mud, which eased the pain. It was risky business, however: one day a seagull stepped on his neck, and the pain was almost more than Newberry could bear. But the way he almost lost his life was by daydreaming one sunny morning when he should have been paying attention to the world around him. He was so busy dreaming that his neck went to sleep. Just at that time, an old rock crab happened by. He was a pretty old crab, and his eyesight was a bit misty. Also, one eye had a bit of seaweed stuck to it. He saw Newberry's neck poking out of the mud and thought it was something good to eat. He didn't know that Newberry was on the other end. So he snapped on to it with his big pincer claw, but since his eyesight was misty, he nearly missed and didn't get a good grip on the neck. Newberry did not feel the crab at first because his neck was asleep, so he did not pull it in immediately. But the sharp pinch finally woke Newberry's neck. It also woke up Newberry from his daydreaming. When he tried to pull in his neck, the pain bothered him so much that he very nearly did not have the courage to yank. But yank he did. The crab was so surprised when his food yanked back that he let go. Newberry's neck came in so fast when the crab let go that he almost gave himself a concussion. That was when he finally decided to see a doctor.

There was a doctor in the town of Blue Hill who had treated many kinds of odd patients. He had treated Freddie Bissel, who cracked his ribs because he thought he could fly

m. 1981

from the roof of his front porch. He had treated old man Wemble when he got his finger stuck in the hole he had drilled in the maple tree during maple sugar time. He had treated Jed Black after he swallowed a mouthful of roofing nails when he slid off the top of Asa Gray's barn during a thick slippery fog. He had even untangled old Stu Fletcher when he darned part of his long whiskers into a pair of socks he was trying to mend. But Newberry was the first clam he had ever been called upon to treat. He had eaten many clams, but he had never treated one. He didn't mention this to Newberry, though, because he thought it might be impolite. Besides, he did not want to admit that there was anyone or anything that he was not smart enough to cure.

"You have a crick in your neck," he told Newberry.

"I know that," said Newberry. "I want you to fix it."

"Well," said the doctor, "I'm not surprised that your neck aches since you spend all your time in a damp mudflat. What you need is a warm mud poultice to draw out the ache."

"I live in a warm mud poultice everytime the tide goes out," said Newberry. "Except," he added, "during rainy weather. This has been an especially wet June."

"Yes," said the doctor, "the worst since 1967. Well, you cannot go around with a pain in your neck. It's quite a remarkably long neck, you know. Quite the longest I've seen in my practice, and I've seen a great many necks. I must certainly report this to the Maine Medical Society. In the meantime, I'm going to prescribe some liniment. Rub it on twice a day, or more often if the pain gets too severe. If the neck is not better in a week, come back and see me."

Newberry went home to his boulder. Faithfully twice each day he rubbed liniment on his neck. It burned a little, but he figured that any liniment with any healing power had to burn a little if it was any good.

The liniment did not make the pain go away, but it did

save Newberry's life. It happened on the Fourth of July, a particularly difficult day. Low tide came early in the afternoon. Several families had picnics on the shore that day. Children were scampering all over the place, so Newberry was constantly yanking his neck in to avoid getting stepped on. All this activity didn't help the pain a bit. At one point, one of the grown-ups sat on the boulder while Newberry had his neck out. Newberry could not move enough to haul it in. He was stuck in that uncomfortable position for fully fifteen minutes. Fortunately, one of the children got into trouble, so the grown-up had to go rescue him.

As soon as he could, Newberry put a double dose of liniment on his poor aching neck. Later in the day, when the picnickers left for Stonington to see the fireworks that would begin as soon as darkness fell, Newberry stretched his neck to get out the kinks—even though the tide had now come in and the water was cold. But he hardly had it out and set in a really comfortable position when a large, particularly hungry clam worm took a tremendous bite of it. How fortunate it was that Newberry had rubbed on a double dose of liniment! The liniment tasted so horrible and burned so fiercely that the clam worm let go instantly. He almost tied himself in a knot trying to get the taste out of his mouth.

Though the liniment saved Newberry's life, it did not make his neck feel the least bit better. When he could stand the pain no longer, he went back to the doctor in Blue Hill.

"You simply have to do something," he said.

"Well," replied the doctor, "I'm really not surprised that the liniment did not help. After all, if you insist on living in that damp mudflat—. Medicine can do only so much. If people refuse to live sensibly, medicine can't help them. Somehow we have to get that neck of yours warm and dry. I suggest that you go home and wrap a pure woollen muffler around it."

"What color woollen muffler?" asked Newberry.

"It really doesn't make much difference," replied the doctor, "as long as it is pure wool."

"I think purple is a pretty color," said Newberry.

"Purple is a very nice color," said the doctor.

"Won't it itch?" asked Newberry.

"By and by you'll get used to it," the doctor said.

"I suppose so," agreed Newberry. Then he thought of something else. "Won't it shrink in the water? It could be very uncomfortable if it shrank while I had it tied around my neck."

"You needn't worry," said the doctor, "as long as you don't let it get into hot water. It would never do to let it get into hot chowder."

Of course it never did because not many people took the trouble to go clamming in rocky mudflats. Newberry became quite a celebrity in his mudflat. The pain in his neck finally went away, but the purple muffler made him look so distinguished that he decided to wear it forever after. It still itches a bit at low tide, but he figures that it is worth it.

2

Perils of a Purple Muffler

Newberry was very happy with his new purple muffler. It was an imported English muffler and one of the best. It not only kept his neck warm, it made him famous among all the animals who lived in the mudflat. At high tide, when the water was deep and it was safe to stick his neck out, Newberry used to let the loose end of his muffler wave gently in the ocean currents like a piece of bright purple seaweed. This wasn't always the wisest thing to do because sometimes a hungry tom cod or a small eel would snap at it, thinking it was good to eat. But now that his neck no longer had a pain in it, he could snap it back in faster than any tom cod or eel could bite. Anyway, the tom cod and the other fish were always so fascinated with the muffler that they aimed at it instead of at Newberry's neck and missed him every time. There was not much to fear from crabs and lobsters because he was quicker than any of them. Sometimes a sea urchin would come and sit on the muffler. Then Newberry would have a difficult time removing him because he held on tight with so many little suction feet.

At low tide things were different. That was when Newberry should have kept his neck in the mud like any intelligent clam, but, I am sorry to say, Newberry had become somewhat of a showoff. He would stick his neck out just far enough for the purple muffler to shine its brightest in the sun. One day a spotted sandpiper stopped to peck at it, but Newberry squirted him in the eye.

The purple muffler got him out of trouble almost as often as it got him into trouble. One scary experience happened one morning when he was dug up by a man who was out clamming with his three children. Newberry had known they were coming by the vibrations in the mud. When the man's shadow fell on his neck, he yanked it down out of sight, but his quickness didn't help him this time. He felt the clam hoe crash through the mud behind him. The next minute, he and a few old neighbors were dredged into the open. "Well, that's the end of me this time," he thought. "That man will surely see me with this bright muffler. How could he miss me? For once I wish I were just like other clams. Then I would be the same color as the mud and perhaps he wouldn't see me."

As the man began picking out the clams from the hoe-full of mud, one of the boys shouted with great excitement. "Look, Dad, there's a clam with a purple muffler!"

"Don't be ridiculous," said the father. "Clams don't wear mufflers, especially purple ones."

"This one does," insisted the boy.

His two brothers came over to look. "He sure does," they said.

Newberry had his neck in as far as it would go, but he never had been able to get it all the way in without having the muffler scrape off against his shells. So there he lay, big as life with the muffler sticking out like a sign saying "Here I am."

The father took his glasses out of his pocket so he could get a better look. "Yes," he said at last, "it certainly looks like a muffler." He thought hard for a moment. "I know, it must be a marker put on by a scientist studying clams. Let's throw him back. There are plenty of others."

"O.K.," said the boys, and one of them tossed Newberry back into the mud.

But Newberry had not learned his lesson. The next morning he stuck his neck out again. The tide was not all the way out, but still it was a dangerous thing to do. The water was cool, exactly the way he wanted it in the early morning. Cool water helped him wake up quickly after sleeping all night. Well, there he was, wide awake, eating his usual breakfast of invisible little animals that the tide swept past his burrow. This morning there was a good current. Besides bringing breakfast it tugged at a loose thread at one corner of the muffler. The bright purple thread waved back and forth and wiggled just like a worm, which is exactly what an old rock crab thought it was. He reached out with his pincer and grabbed the loose thread. Of course, when he got the thread into his mouth he realized that it was no worm. Naturally, he tried to spit it out, but fuzzy woollen threads have a habit of sticking to things, especially things like spiny claws and crabs' mouths.

When the rock crab discovered that he couldn't let go of the thread, he crawled off sideways and backwards trying to pull free. Ordinarily this would have worked fine, but this time it didn't work at all because as he backed away the thread began to unravel from Newberry's muffler. In the meantime the tide had run out all the way. What a sight there was when the whole mudflat came into view! There for everybody to see lay a long purple thread with an old rock crab at one end and Newberry at the other. The crab was trying desperately to get untangled from the thread while Newberry, with his neck stretched to the limit, was trying just as desperately to wind the thread back in around his neck before the muffler unraveled completely and disappeared.

While this was going on, a herring gull, who had not had any luck at all in finding breakfast, glided over the mudflat. The first thing he saw was Newberry's neck. He landed a few feet away and walked over carefully to inspect this strange object. You must remember that he had never before seen a clam with a muffler, so he was being pretty cautious. When he arrived closer, he cocked his head to one side for a more careful look. Newberry was too busy trying to wind up the thread to notice the gull. When the gull saw the neck move, he decided that the strange object was good to eat. He grabbed the neck, yanked the surprised Newberry out of the mud, and swallowed him muffler, shell, and all, in one mighty gulp.

It is hard to say who was more surprised, Newberry at finding himself in a herring gull's stomach, or the herring gull at finding that he had swallowed a whole clam. If he had realized that the strange object was part of a clam, he would have done what he always did: instead of gulping it down he would have held the clam in his beak, flown straight up into the air, and dropped it on the rocks below to smash it. Instead, here he was with a big, hard clam in his stomach. The lump was so distressing that he did not even notice that a purple thread hung out of the corner of his mouth. He felt pretty uncomfortable so he threw back his head and made the most mournful herring gull cries ever to tell the whole world how uncomfortable he did feel.

Another herring gull who was also out searching for

breakfast heard the unhappy cries. She flew over to investigate. She glided over the mudflat and came down to land just a few feet away from the first gull. Before the first one could explain his trouble the second one discovered the old crab, who was still trying to untangle himself from the thread of Newberry's muffler. In the twinkling of an eye, the old crab disappeared into the second herring gull's stomach.

Seeing nothing else to eat, the second gull jumped into the air, flapped her wings, and began to fly out to sea. She had not gone more than three feet when something tugged at her mouth and jerked her back down to the mudflat. At the same time, the first gull with Newberry inside was also jerked off his feet. He fell face down in the mud and his neat white coat became covered with oozy gray stuff. The two gulls looked at each other. It took them only a minute to realize what was wrong—they were hitched together by the thread of the muffler. One end of the thread was attached to Newberry, who was in the first herring gull's stomach, while the other end was attached to the rock crab, who was in the second herring gull's stomach. Neither gull wanted to give up his breakfast.

Then began the most ridiculous tug-of-war imaginable. For a while it looked as if the thread would break and both gulls would keep their breakfasts. Fortunately for Newberry, the muffler had been woven from the finest wool so the thread was exceedingly strong.

First one gull, then the other would try to fly away. Each time he would be hauled down out of the air. This yanking and pulling was very upsetting on stomachs. It was also upsetting on old Newberry and the rock crab. Finally, the first herring gull could stand the strain no longer; no breakfast was worth this much trouble. At the next yank he opened his mouth wide. Newberry came flying out of his stomach, landing on the mud with a squashy plop. At the same time, the second gull suc-

ceeded in biting through her end of the thread and flew away with the poor old crab still in her stomach.

Newberry was so happy at being free that he wasted no time in burrowing out of sight in the mud. He did not even bother to reel in the loose end of thread. There would be plenty of time to take care of that later. Anyway, he planned to cut the extra thread loose on the sharp edge of his shells. His handsome purple muffler would be shorter than before, but it would be neater without the loose ends—and a great deal safer.

3

The Two Day Tug-of-War

When Newberry woke up on the Tuesday after the Fourth of July, he did not dream that he was headed for trouble. The trouble really got started a week earlier, but Newberry did not know that either. It all started when Mort the lobsterman hauled up a trap set near Newberry's island.

Mort followed the same routine with every trap. First he picked out the good lobsters and crabs. Then he threw overboard the lobsters that were too short. After that he threw overboard the old bait that the seagulls were crying for and replaced it with fresh bait that smelled just as bad. The last thing he did before pushing the trap back into the ocean was clear out the starfishes and sea urchins that clung to the slats. He threw them overboard too.

Mort wasn't too careful with starfishes and sea urchins because he didn't like them. They messed up his traps and also his boat. So, many times the sea urchins got smashed like eggs, and the starfishes lost a leg or two. On this particular day Mort accidentally cut off the leg of one large starfish.

The slippery, smelly mess of crushed sea urchins, whole sea urchins, sea urchins that were bald because all of their spines had rubbed off, whole starfishes, half starfishes, the starfish with the missing leg, and the leg itself, splashed overboard. Everything sank helter skelter to the bottom before the seagulls could snap up even the smallest bit.

Down on the bottom, crabs and fishes rushed about fighting over the pieces and gobbling up every scrap. Even starfishes arrived. They didn't rush, but they got there in time to pick up a few crumbs. Some of these crumbs were sea urchins and starfish crumbs. Starfishes didn't mind eating their relatives. Even the starfish with the missing leg hobbled over to the feast. The next day he got an infection and died, but the leg, which had fallen into a crack in the rocks where nobody could fish it out, remained very healthy.

A few days later, ocean currents carried the leg into Newberry's mudflat, where it drifted around and soon began to get very hungry. Now what can be worse than to be a healthy, hungry leg with no mouth and no stomach? The leg had no brain either so it couldn't think about these things. But it could grow, and grow it did. It grew into a brand new whole starfish complete with mouth, stomach, and five little legs. The stomach was no bigger than the leg from which it had grown. Just the same, it had a huge appetite. It had an appetite bigger than itself.

At last it could hunt for something to eat. What it liked best were clams and mussels. It really should have started with a small clam or mussel—one its own size. Instead, it tried to eat the first clam it found. The first clam was Newberry.

Usually Newberry remained safe down in his burrow. However, earlier that day a clam digger had dug up the mudflat so thoroughly that the high tide had washed Newberry and many of his neighbors out into the open. The starfish found him before he had time to dig back into the mud.

As soon as Newberry noticed the starfish he pulled in his neck and closed his shells as tight as a stuck door. But the starfish had a way of opening stuck doors, especially if they were clam shells. First he climbed on top of Newberry, who could hear the little suction feet go *squish* and *pop* like raindrops on a roof. Then he grabbed Newberry's left-hand shell with the suckers of two of his legs and Newberry's right-hand shell with all of the suckers on two more of his legs. He didn't quite know what to do with leg number five, which was neither on the right nor on the left.

While he was trying to make up his mind, which never worked very fast anyway, a loose end of Newberry's purple muffler, which stuck out between the two shells, got in the way and tickled the starfish. After a few tries, the starfish caught hold of the muffler to keep it out of the way. That stopped the tickling and gave him an idea as well. He would pull Newberry out of his shells by the muffler.

At first nothing happened. A few threads came loose and drifted away. But as the starfish pulled and pulled the knot in the muffler got tighter and tighter around Newberry's neck till Newberry began to choke. Now, when Newberry chokes he squirts water out through his neck. That is exactly what he did this time. A mighty squirt caused the starfish to let go of the muffler, and he almost lost his balance.

The starfish decided then to do things the right way, the way starfishes were supposed to do things. He would simply pull

M. 1981

Newberry's shells apart. So began the great tug-of-war. The starfish pulled from the outside; Newberry pulled from the inside.

Now, neither Newberry nor the starfish ever moved very fast, but they both were very stubborn. The great tug-of-war began on Tuesday morning. By Tuesday afternoon the two were still pulling, and neither had gained an inch. They pulled all afternoon and all night. Wednesday morning came and they were still pulling. The starfish was pulling for his dinner. Newberry was pulling for his life. If this had been a full grown starfish it might have been the end of poor Newberry. It was still nip and tuck because the starfish had so many little suckers on each leg that it could rest some while the others kept up the pull. Newberry had only two big muscles, and these were beginning to get a bit tired.

The old barnacle on the rock near Newberry's burrow watched the struggle with great interest. "How are things going?" he asked Newberry, but Newberry was much too busy for conversation so only a grunt came out from inside his shell.

"I'll get him," said the starfish.

"I didn't ask *you*," replied the barnacle. "Why don't you young starfishes tackle clams your own age and respect your elders?"

"I'm not a young starfish," said the starfish, "I'm an old starfish."

"How can you be so old when you're so small?"

"I grew from one of my legs."

"That's ridiculous," snapped the barnacle. "Nobody can grow from his own leg."

"Well *I* can, and I'm much too busy to try to explain it to anybody as dull as you," the starfish replied impolitely.

That insult hurt the barnacle. It made him so angry that he snapped his shell shut and sulked.

The two had been so busy trading insults that they did not see a small eel no bigger than Newberry himself that had swum over to watch the tug-of-war. At that moment, the starfish managed to pull Newberry's shells apart just a crack. Now, most of us eat by putting food in our stomachs, but starfishes eat by putting their stomachs in the food and digesting it inside out. As soon as the starfish discovered the small crack, he turned his stomach inside out and dropped it into Newberry. This was too much for Newberry.

"No fair tickling," he grunted. "Keep your stomach to yourself."

Just then the tickling made him sneeze so hard that his shells snapped together and gave the starfish's stomach a mighty pinch.

"Ouch!" cried the starfish as he pulled his stomach back where it belonged.

The eel had never seen anything so funny. He swam closer for a better look, but he came too close. He bumped into the rough coat of the starfish, which had a lot of little pincers. The pinch from the starfish's coat caused the eel to wriggle and twist so that he almost tied himself in a knot. He tried to rub the itchy spot on the nearest thing at hand. The nearest thing was Newberry. He rubbed all over Newberry, and rubbed off a lot of the slime that made him slippery as an eel. Everything became a slimy mess.

All of a sudden the starfish began to slip in the slime. First one leg slipped, then another, then all five at once. Waving wildly, he slid off Newberry and tumbled slowly to the bottom. Before he could clean himself or get a hold on something, anything at all, a trickly current picked him up and tumbled him in big somersaults out to sea. At the last moment he tried to grab a branch of seaweed to save himself, but a middle-sized fish, a cunner whose eyesight was not especially sharp, smelled the eel slime, thought that the starfish was a delicious young eel, and snapped him up in one gulp.

Back in the mudflat a very tired Newberry relaxed his muscles, pushed out his neck, which was very cramped, and gave a great sigh of relief that ended up as a gurgle as he sucked in a huge gulp of cool water.

"That was a close one," said the barnacle, who had gotten over his sulk.

"Much too close," agreed Newberry and promptly went to sleep while his purple muffler waved gaily in the incoming tide.

4

Newberry's Birthday Feast

August fourteenth was Newberry's birthday. He was five years old, which is pretty old for a clam. It was like being an old, old man, but Newberry did not feel old, and he wanted to celebrate his birthday. How can anyone celebrate a birthday in a mudflat? He couldn't bake a cake, and even if he could there was no way to light candles. Newberry decided instead that he would have a great banquet. He would eat and eat until he couldn't hold any more. He would stuff himself as he had never stuffed himself before. So when the tide started in, Newberry began to eat all the delicious tidbits that the currents brought to him. He didn't have to stir from his doorstep; the tide surging in over the flat brought all sorts of special things. There were the eggs of sea animals so small that he couldn't even see them. There were the tiny animals themselves. He sucked them all in. For this main course he ate little floating animals and larvae. For his salad he stuffed himself with the tiniest of plants.

Before he quite realized what had happened, he had stuffed himself so full that there was hardly any room for him between the shells. Even his burrow was too tight. He pulled himself up near the entrance. He felt mighty uncomfortable. Then the trouble really began—he got terrible indigestion. A big bubble of gas formed in his stomach. Now it is bad enough to have a bubble of gas in one's stomach because it hurts, but in Newberry's case the trouble was worse. The bubble made him so light that before he knew what was happening he floated right up out of his burrow to the top of the water. There was an offshore breeze—it began to blow him out to sea.

If the bubble had not been so painful, Newberry might have enjoyed himself. He was a venturesome soul and liked to travel, but being a clam he had never had much opportunity. Now here he was sailing grandly out to sea his bright purple muffler trailing in his wake. What a stirring sight! What a pity that he was not really enjoying it.

Just as Newberry was rounding the ledge on the east side of the mudflat and got caught in the rip tide between the two islands, a great black-backed gull appeared on his starboard bow. This time it surely seemed that Newberry's end was in sight. The gull spotted him bobbing along on the tide. He was so conspicuous with the long end of the purple muffler trailing out behind that any gull would have to be blind not to see him. Fortunately for Newberry, Mort the lobsterman hove into sight at just his moment. Mort had eighty traps out, each one marked with a bright white and yellow buoy. In his boat he had a big barrel of pickled redfish for bait. As he hauled each trap in he would take out the lobsters, throw back the short ones and the ones with eggs, then pick out the crabs, which he threw in a bucket, and lastly, take the old bait out of the bait bag and fling it overboard. This was just what the great black-backed gull relished. He couldn't be bothered with one sick clam. He would do better to follow Mort's boat. So he wheeled away, leaving Newberry to bob safely but helplessly before the wind.

By two o'clock in the afternoon he had drifted past the red nun buoy marking the channel into Blue Hill. A cormorant sitting on top of the buoy with his wings stretched out to dry looked down at him.

"You're on the wrong side," he said.

"What difference does it make whether I'm on my right side or my left side?" said Newberry. "One side hurts as much as the other."

"You're on the wrong side of the buoy," said the cormorant. "Leaving the harbor you are supposed to have the red buoy on the port side."

"Right now," said Newberry miserably, "I wish I were entering the harbor instead of leaving it. What will become of me if I go out to sea?"

"Someone will probably eat you," the cormorant remarked, "but then, I'm not so sure. You look silly with that purple muffler trailing out behind."

Newberry didn't hear the cormorant answer because a puff of wind had blown him out of hearing. There is no telling how far out to sea he might have been blown if the great black-backed gull had not stopped following Mort's boat because he was tired of eating old bait. He sailed around looking for something for dessert. That was when he spied Newberry. He swooped down and picked poor old Newberry neatly off the crest of a curling wave.

Without missing a wing beat he turned back toward the island with the mudflat. He was searching for a rock on which to

m. 198

crack open his dessert. Finding one ledge that looked particularly nice, he hovered over one end of it, and opened his beak. Down went Newberry, his purple muffler flapping wildly as he fell. He was certain that his end had come, but instead of landing on the rock and smashing into a millon pieces he landed on a big patch of seaweed. The seaweed was thick and slippery and full of little bladders of air which acted as a cushion. Even so, he landed with a tremendous thump that almost knocked the bubble out of him. His troubles were not over, however, because the gull had followed him all the way down. Before Newberry could even wonder where he was, the gull picked him up, flew straight up in the sky, took aim on a bare spot on the ledge, and let go. Once more Newberry tumbled toward the ground. The seagull's aim was pretty poor, and again Newberry landed in seaweed. "I can't stand much more of this," he thought as his insides sloshed around after the tremendous bump. Three more times the gull dropped him. Each time he missed the rock.

"I'll try once more," said the gull to himself. He picked up battered and bruised Newberry, flapped up, up, and up, took extra careful aim and let go for the last time. Down tumbled Newberry, twisting and turning in the air until he was quite dizzy. When he was halfway down a gust of wind, stronger than the rest, swept in from the bay. It capsized one of the racing sailboats, blew the cormorant off the buoy, flipped the seagull into a spin, and blew Newberry three degrees to the east. He landed on a big patch of seaweed at the very edge of the ledge, bounced once, and slid into the water.

The gull was disgusted. "Well," he said to himself, "he was probably tough. Besides I really don't feel like having dessert. I think I'll fly over to the other side of the island and see what that family on the shore is eating at their picnic."

At last Newberry was left alone to bob up and down in the waves. The waves were getting bigger. There were even a few

m. 81

small whitecaps. There was no doubt about it—the sea was getting rougher. Before long, Newberry became horribly seasick. "What a day," he thought, "I've never felt so miserable. My stomach aches, my insides are all scrambled, my brain is bruised, and now I feel as though I am going to lose my dinner!" He turned pea green. He was not used to bobbing about on top of the water. Down on the bottom where he lived the sea was always calm. The mudflat was a good solid place. It never heaved up and down. He tried very hard to keep from getting sick. He tried to think of pleasant things.

It was no use. His stomach just would not listen to his brain. It growled and grumbled and bubbled and boiled. He couldn't hold his dinner any longer. Up came that wonderful birthday dinner—every last bit of it. At the same time he felt a tremendous burp. Up came the bubble with the dinner.

No sooner was the bubble out of Newberry than he lost all of his buoyancy. Without the big bubble inside of him he no longer floated. Slowly he began to sink. The sea became calmer. He slid smoothly toward the bottom and landed gently in the mud. The currents of the tide tumbled him softly along and finally bumped him gently against his very own rock. With a great sigh of thankfulness, Newberry wasted no time in burrowing deep into the mud, where he promptly went to sleep and didn't wake up for three days and nights.

5

The Unwelcome Guest

It had been such a warm night that Newberry had slept, as he often did, with his neck part way out of the burrow and his purple muffler waving in the currents. It was a foolish thing to do; when he awoke he had to clean a whole teaspoonful of sand grains out of the muffler. (He was a tidy clam.) He also noticed that there was a very small blue mussel on the muffler.

There had been no mussel on the muffler the day before. Newberry was certain of that because he was, as I have said, a very neat clam. The mussel must have arrived during the night and decided that the muffler was a splendid place on which to make a home. He probably did not know that the muffler belonged to somebody. Besides, mussels are not very particular about whose belongings they choose for setting up housekeeping.

This mussel was very small indeed and probably was not much of an inconvenience. But small mussels grow into large mussels, and a large mussel hanging around one's neck *would* be a nuisance.

Newberry had no particular liking for mussels anyway, even though they were distant cousins. They cluttered up the mudflat so that other folks could not live comfortably. As soon as one mussel moved into the mudflat, others followed. One or two here and there would have been fine. They would have added to the interest of the mudflat life. Hundreds would be impossible.

Mussels lived side by side so closely that they covered every square inch of mud so that clams and feathery fan worms had no room to burrow. How could one dig a burrow through a wide blue carpet of hard shelled mussels? Mussels also ate a tremendous quantity of food so that there was not much left for other folks. No, Newberry had no love at all for mussels.

When he discovered the wee mussel on his muffler, he lost his temper. That mussel had a lot of nerve setting up housekeeping on somebody! When Newberry lost his temper, it was a terrible thing to see. He couldn't turn red and shout and stamp his feet, or go into a corner to sulk, as people do, but there was no doubt that he had lost his temper. He squirmed and squirted and ground his shells together with an unpleasant gravelly sound. The barnacle who lived on the boulder nearby closed his own shell to keep out the noise, and a clam worm who was thinking just then about having a fresh Newberry supper decided that Newberry was so unpleasant that he would probably taste sour. Meanwhile, Newberry swished his muffler violently through the water to shake off his unwelcome guest.

But the mussel stuck as though he had been glued there,

and glued he really was. He had made up his mind to stay. While Newberry was sleeping, the mussel had spent a busy night spinning small but very strong threads of glue from his shell to the muffler. Now he was tired and wanted to rest. He was anchored forever no matter how much Newberry might wriggle and squirm and thrash.

When none of Newberry's antics succeeded, he tried to scrape off the mussel by pulling his neck in and out of his shell. That trick did not work at all, mostly because the muffler kept coming undone. Also his neck was getting sore from all the scraping back and forth. Newberry was desperate.

While he was resting and wondering what to do next, a gray shrimp came crawling by in search of some breakfast. He stopped to look at Newberry.

"What in the world are you doing?" he finally said

"I'm trying to get rid of a mussel on my neck," replied Newberry.

"You mean *in* your neck,' said the shrimp.

"No, I mean *on* my neck, just as I said," said Newberry irritably.

The shrimp had never heard of anything so ridiculous. "Why would anyone want to get rid of one of his muscles?"

"It's not *my* muscle. It's not anybody's muscle. It's a blue mussel, and he seems to think that my muffler is a good place to make his home."

"Oh," said the shrimp, "you can't blame me for thinking about the wrong kind of muscle. After all, who ever gets the other kind growing on him?"

"*I* do," said Newberry sharply.

"Well I shouldn't wonder. What can you expect if you lie around day and night like an old rock. I'm surprised that you don't get completely covered with seaweed and barnacles and all sorts of things. One has to keep moving in this world.

Newberry was becoming more angry all the time. He didn't like shrimp. He didn't like this particular shrimp. It was cross-eyed and mean looking.

"Until today, I've lived quite happily here without gallivanting all over the place."

The shrimp paid no attention to this impolite remark, but said, "It's your own fault anyway for wearing that old blue rag around your neck."

"It's a muffler," said Newberry, "and it's purple, not blue, and it was very expensive." Then he had a thought, "How is it that you do not have any mussels growing on you?"

"First," replied the shrimp, "my shell is smooth and shiny, not rough like yours."

Newberry kept quiet.

"Second," added the shrimp, "I keep moving about all the time."

Newberry still kept quiet.

"And last," said the shrimp, "I don't wear old mufflers."

Newberry was too tired to argue any further. "But if a mussel *should* start growing on you, how would you get rid of it?"

"I'd lose it when I shed my shell," said the shrimp.

"When you *what*?" cried Newberry.

"When I shed my shell."

Newberry had never heard of anything so queer, "You mean you just take off your shell and stand around with nothing on?"

"Of course not," snapped the shrimp. "I grow a new one underneath before I take off the old one." And then he looked at Newberry with new interest. "Don't you grow a new shell?"

"No," Newberry replied.

"Then how do you make room for yourself in the old one when you grow?"

"I just keep adding new shell to the edge to make more space inside."

"The edge?" exclaimed the shrimp, puzzled. "I haven't got edges."

This was a whole new mystery. "I thought everything had edges," Newberry said.

"Look at the blueberry that the tide washed up over there," and the shrimp pointing one of his claws in the proper direction. "It hasn't any edges."

"Well I have," said Newberry firmly, "and it's a much more sensible way of growing than throwing away a perfectly good shell when it gets too tight. That's very wasteful."

But the shrimp lost interest in the argument. He didn't

understand what edges were, and he was hungry. He was looking very attentively at the mussel on Newberry's muffler.

"I can get rid of the mussel for you," he said at last. "Like this!" And before Newberry could say yes, no, please, or thank you, the shrimp had reached out with his pincer claw, picked off the mussel, and popped it in his mouth. There was a horrible crunching noise, and the mussel disappeared, shell and all.

"Thank you for the breakfast," said the shrimp and waved gaily with his long feelers as he crawled off in search of dessert. "The next time you get covered with mussels let me know. I'll come by and tidy you up a bit."

Newberry couldn't think of a thing to say.

6

A Message in a Bottle

When the tide went out one Tuesday morning, it left a bottle sitting in the mud almost on top of Newberry. Newberry was annoyed. The tide was always leaving things on the mudflat, but usually they were interesting bits of driftwood or Irish moss, or a stranded jellyfish. Newberry hated it when the tide dumped trash in his front yard. People on boats had no regard for front yards when they threw trash overboard.

This bottle looked different, however. It seemed to have something in it. The something appeared to be a note. Newberry was immensely curious about that something in the bottle. Was it really a note? And if it was a note, what did it say? Fortunately, the cork in the bottle was quite rotted. The bottle must have been in the sea a long time. It was no trouble at all for Newberry to push what was left of the cork down into the bottle. He stuck his neck into the bottle. He intended to suck the note out and read it, but he suddenly remembered that he had never been to school and never learned to read. Bitterly disappointed, he started to pull out his neck. It wouldn't come. He was stuck. You see, when he had stretched his neck to reach into the bottle it had become thinner and thinner as he had stretched it, so getting in was easy. When he started to pull his neck out, it got thicker and thicker as it got shorter. It simply would not budge. The harder he pulled the more tightly he got stuck. There was nothing to do but go once again to see the doctor in Blue Hill. Fortunately, the bottle was a small one.

When the doctor saw Newberry, he realized immediately what the trouble was. "How did you get into that predicament?" he asked.

Newberry started to answer.

"I can't hear you," said the doctor. "The sound is all bottled up."

Newberry tried again.

"Wait a minute," the doctor said. "Perhaps if I put my stethoscope to the bottle I can hear you. Oh, yes. That's much better. You have no idea how funny you sound talking in a bottle. It has a booming sound. Reminds me of the surf at Schoodic Point."

"Get me out of here!" boomed Newberry.

"Ah, yes, of course," said the doctor. "Now let me see." He picked Newberry up in his left hand and the bottle in his right and tried gently to pull them apart.

"Ouch!" boomed Newberry.

"I can see that that won't work," said the doctor. "Perhaps if I place the bottle in hot water the air will expand and blow you out."

m. 1981

"Oh, don't do that!" cried Newberry in great alarm. "I cannot stand much heat. I'll be steamed if you heat the bottle."

"Steamed clams are delicious," the doctor said.

"You shouldn't say things like that," said Newberry sadly.

The doctor thought for a while. "Hmm, if I could soap your neck, it might just slide out by itself." He got some special hospital soap and rubbed it carefully around the neck just where the purple muffler ended and the edge of the bottle began. But the soap did not help a bit because none of it got down into the neck of the bottle where the trouble really began.

"I have an idea," boomed Newberry. "Why don't you break the bottle?"

"That's very dangerous," the doctor protested. "You might get cut."

"Please try," said Newberry. "I would rather have a few cuts than be stuck here the rest of my life."

"Well," the doctor hesitated. "Let me see what kind of a bottle this is." He looked at it carefully. It was some kind of old medicine bottle. The raised letters in the glass had been smoothed down by the rubbing of waves and sand; however, the doctor was still able to read them. What they said was FROG POND AND FEVER REMEDY.

The doctor became terribly excited. "Why," he exclaimed, "this is a very, very old medicine bottle. It is an antique. Wait until I get my catalogue on antique bottles." One of the doctor's hobbies was collecting old medicine bottles. He had a fine collection of them. As a matter of fact, his collection was the best in New England. Each fall he exhibited some of the rarest ones at the Antique Show in Ellsworth.

He returned in a minute with the catalogue that he always kept handy on the shelf over his desk between the pharmacopoeia, which told all about drugs and poisons, and *Gray's Anat-*

omy, a big fat book with pictures of all parts of the body. Quickly he thumbed through the pages.

"Ah, here it is. Frog Pond and Fever Remedy. Very Rare. Value thirty-two dollars."

"I can't break this bottle," he told Newberry. "It's very old and rare. It is quite valuable."

"But," protested Newberry, "I am very valuable to me and my neck is becoming pretty rare from the rubbing it is receiving in this old Fish Pond bottle."

"Frog Pond," corrected the doctor. "Well, I have one more idea." He got one of his hypodermic syringes and filled it with soapy water. "Now," he said, "do not wiggle your neck." Very gently he pushed the needle between Newberry's neck and the neck of the bottle.

Newberry wiggled.

"Stop it," said the doctor.

"It tickles." replied Newberry.

"I don't care if it tickles. If you wiggle, you will get a neckful of soapy water, and that will smart."

Newberry kept as still as he could although it was very difficult. Slowly the doctor worked the needle all the way down into the neck of the bottle. Then he began to pull it out slowly, and as he did so he kept squirting out a little bit of soapy water. When that was done, he again picked up Newberry in his left hand and the bottle in his right, and pulled. Newberry's neck slid out of the bottle as smooth as smelt and with a little pop.

"There," said the doctor, "you're out."

"Thank you," replied Newberry politely.

"You are quite welcome. Please be careful where you stick your neck next time."

Newberry promised that he would. "I wouldn't like to be covered with that soap again," he said. "It tastes too much like medicine."

"It is not supposed to be tasted," the doctor told him.

"Oh, by the way," said Newberry, "I almost forgot. There is a note in that bottle. Would you be so kind as to tell me what it says?"

The doctor took a long pair of tweezers from his black bag. He poked down into the bottle. At first he could not manage to get a good grip on the note because the paper had become softened from Newberry squirting all over it every time the doctor tickled his neck, but by being very patient he managed to coax the note out of the bottle. He smoothed it out very carefully. Even then he had some difficulty making out what it said because Newberry's squirts had smudged the ink. At last he said, "It's nothing important."

"It must have been important to be put in a bottle," Newberry remarked.

"Oh, I think somebody just stuffed it into the bottle to get it out of the way and somehow the bottle got thrown out with the trash."

"Please tell me."

"Well, if you must know," the doctor replied, "it is an old recipe for New England Clam Chowder."

7

Newberry's Rainbow

One warm sunny day in August about four o'clock in the afternoon, the tide had just gone out below Newberry's burrow. He stretched his neck out in the warm sun. It felt so good after six hours in the cold mud. Then Newberry heard a wonderful sound. All the crickets and grasshoppers in the marsh grass along the edge of the mudflat were singing their happiest songs. He had never heard such a wonderful sound. But then, his memory was not too good. A fellow tends to forget a lot during the long cold winter months when his mind is in the deep freeze. He had heard the grasshoppers and crickets every August since he had been born. Just the same, he lay stretched out enjoying the beautiful chorus as though he really were hearing it for the first time.

m. 1981

I wish I could sing like that, he thought to himself. He mentioned this to a barnacle who lived upside down on the part of the rock overhanging his burrow. After a long pause, while Newberry thought the barnacle had fallen asleep in the sun, the barnacle said, "I've not heard much singing in the mudflat. Some of the shrimps make a racket at high tide, but you would hardly call it singing. And a few fish make nasty sounds."

"I wonder how they do it," said Newberry.

"With their mouths, I guess," said the barnacle.

"No, I mean the crickets and the grasshoppers," said Newberry.

"I guess you will have to ask one of them," the barnacle replied.

At that moment, just by luck, a meadow mouse scurrying through the marsh grass scared the crickets. They scattered in all directions without paying any attention to where they were going. One little, shiny, black fellow made a tremendous leap and landed with a wee splash almost on top of Newberry's neck.

"I beg your pardon, but—," began Newberry.

"Oh, I beg *your* pardon," interrupted the cricket politely. "I didn't mean to land on your nose."

"That's quite all right," said Newberry. "It's not my nose, and you are very light although your feet do tickle."

The cricket hopped off Newberry's neck and onto the sand.

"I've been enjoying your singing," remarked Newberry pleasantly.

"It *is* fun," the cricket agreed.

"If you don't mind my asking, just how do you do it?"

"Oh, it's very simple. I just rub my wings together this way." He made a lovely tickling trill.

"I wish I could do that," sighed Newberry.

"Give it a try," the cricket said cheerfully.

"But I haven't any wings."

The cricket looked embarrassed. "Maybe you could do it the way some of the grasshoppers do."

Newberry looked eager. "How is that?"

"They rub their legs against their wings."

"I've got a foot," said Newberry doubtfully. He pulled it out of the mud.

The cricket examined it critically. "I am afraid that won't do it. I've got an idea. Why not try rubbing your shells together? They are just like great big wings. They should make a tremendous song."

Newberry beamed. He flexed all his muscles and rubbed his shells back and forth. There was a horrible grating sound like somebody trying to cut off the edge of his dinner plate with his knife. The cricket shuddered.

"Perhaps there is a sand grain stuck in there someplace," he said sympathetically.

Newberry sucked in a great gulp of water from a convenient puddle in the mud, sloshed it around inside himself like a small boy rinsing out his mouth, then squirted it out hard. He had completely forgotten where the cricket was. He washed the poor fellow right off the pebble on which he was perched.

"Oh, I'm terribly sorry," apologized Newberry. "I forgot that you were so close."

"Accidents will happen," said the cricket cheerfully as he spit out a few grains of sand. "Try it now."

Newberry made up a beautiful song in his mind, counted three, and rubbed his shells together. His beautiful song came out: "SCRATCH, GRIND, RATTLE."

The cricket shuddered again. "Your hinge is rusty," he said.

"My hinge does not rust," said Newberry with some irritation. "I am not made of iron, you know."

The cricket looked interested. "What *are* you made of?"

"My shell is mostly calcium. That's a mineral from the sea. It never rusts."

The cricket looked impressed. "Just the same, it is squeaky. Perhaps you should oil it."

Newberry tried to change the subject. "Come to think of it, your song sounds rather squeaky." Now it was the cricket's turn to feel irritated.

"My song," he said with great dignity, "has a beautiful pure tone."

"I am sorry," apologized Newberry. "I do think it is nice." He sighed. "I don't believe that I can sing by rubbing my shells together." He sounded terribly disappointed. "It takes practice," the cricket said cheerfully. "Work on it for a few days. I'm afraid I must go now. They will begin to miss me in the chorus. Don't forget—practice, practice, practice."

For three days at each low tide Newberry practiced. Finally the barnacle could stand the squeaks and scratches no longer. "I don't mean to complain," he told Newberry, "but I think your singing is beginning to loosen me from this rock."

Newberry admitted that he was not making progress. "I am afraid that clam shells are just not built for singing."

"Why don't you try singing like a bird?," the barnacle suggested.

"How does a bird do it?"

"I'm not entirely sure. I suggest that you ask the spotted sandpiper the next time she runs by this way. Just between you and me and the seaweeds, she is no great shakes as a singer but she is a bird, and she might know how other birds do it."

This struck Newberry as a sensible idea, so at each low tide he waited patiently for the sandpiper to come his way. He knew that sooner or later she would because she hunted over the flat pretty thoroughly, especially now that she had three young ones to feed.

Again luck was with Newberry because less than a week passed before the sandpiper came his way.

"Excuse me," he called, "but could you spare a moment to tell me how birds sing?"

"We blow air out of our mouths," the sandpiper explained. "Can you do that?"

Newberry felt cheerful again. "I'll try."

He stuck his neck up out of the mud and sucked in a great clamful of air. It was a most unpleasant feeling because the air went down into his insides where only water was supposed to be. It tickled and felt very dry. His insides began to pucker. But he was determined, and he was brave. He pumped himself full of air.

When he could get no more inside, he began to blow it out. It made a most impolite sound. He blew more slowly. Nothing came out but a sigh and a few burps when the air became mixed with some of the water still left inside him. He tried blowing harder. The air came out with a swish. That would not do either. No matter what he did no song came out of him. His insides felt so dry and itchy and puckery that he had to suck in a great clamful of water to keep from sneezing. How good and smooth and juicy it felt!

"It is no use," he told the sandpiper. "My insides are made for water, not air."

"In that case," said the sandpiper, "perhaps you can sing by blowing out water instead of air."

Newberry was doubtful. "Does anybody sing with water?" The sandpiper scratched her head with her foot. "I really do not know, but it is worth a try."

So Newberry tried. The result was most disappointing. The water came out in a beautiful current—absolutely silent.

"Look!" piped the sandpiper excitedly. "How gorgeous! How lovely! You are an artist!"

"Look at what?" asked Newberry.

"Why your squirt, of course. It is making a beautiful rainbow in the sun."

And indeed it was.

"How I wish I could do that," the sandpiper said. Then she had to run away because a small wave came rushing up the mudflat.

Newberry hardly noticed. He was so excited that he forgot all about singing. He sucked in another large clamful of water to squirt another rainbow. Before the new tide came in he made seven rainbows. He could hardly wait for tomorrow.

8

Mysteries

On especially warm sunny days Newberry's island became a very lively place. The birds sang their most beautiful songs, the big, fuzzy, black and yellow bumble bees bustled around the wild roses and butterflies sailed back and forth in the light breezes. One fine summer morning a tiger swallowtail even landed in front of Newberry. She uncoiled her tongue and began sucking up salt water.

This might seem strange to some people, but it seemed quite all right to Newberry, who thought that salt water was the only real kind of water. He had once drunk some rain drops, and they had a horrible taste. As far as Newberry was concerned, anybody who drank fresh water must be crazy.

He watched the butterfly for a while, then squirmed

around to see what else was happening. Out on the water several boatloads of people were coming to the island for a picnic. Yes, it would be a lively day. One family from Surry had already arrived and was exploring and picking up odds and ends near the high tide mark. The little girl of the family found a large moon shell which she held against her ear.

This seemed strange to Newberry. He had seen many animals, even people, put things in their mouths, but he had never seen anybody trying to put something in his ear. "She'll never get it in," thought Newberry, "it's too big."

Suddenly, in great excitement the little girl ran to her mother. Newberry could not hear what she was saying, but it must have been something extraordinarily important because the whole family gathered around and took turns holding the moon shell to their ears. What in the world could they be doing? Surely they were not trying to eat it!

While Newberry was puzzling over this strange behavior, a sandpiper came running across the mudflat, stopping here and there to peck at a slow worm or a careless shrimp. As soon as she was close enough, Newberry asked her if she knew what was happening.

"What are those people up there doing?"

"I've been told," replied the sandpiper, "that when they hold shells to their ears they can hear the ocean roaring."

"That's odd," said Newberry, "the ocean is as flat as a tide pool this morning."

"So it is," nodded the sandpiper.

"I don't understand," said Newberry in a puzzled tone.

"Oh, don't worry about it. People do odd things all the time. Just the other day I saw one of them eat a raw clam."

Newberry shuddered. "That's unkind," he said.

"I'm sorry," apologized the sandpiper, "I wasn't thinking. It's just that everyone who comes out here seems to burn things

in a fire before eating them. I guess *that's* the odd thing—eating burned food."

Newberry tried to bring the conversation back to less painful things. He began another question. "I wonder . . ." but the sandpiper was already running up the beach in search of sand fleas. Newberry ended up talking to himself. He always talked very slowly just after the tide went out because the cold water slowed him down, and it took a while to warm up his conversation. The sandpiper had been in too much of a hurry to wait for each word to come out.

"I'm sorry," she piped back over her shoulder, "I'm in a hurry; I have to get back to my nest to sit on some eggs."

Newberry became more puzzled than ever. People listened to the ocean in empty shells; sandpipers sat on eggs. What a queer place the world was!

Newberry's thoughts were all mixed up. He began to wonder why anybody would sit on eggs, and why the eggs did not break or squish. Clams had eggs,tiny balls of jelly without shells, but nobody ever sat on clam's eggs. That would have made a horrible mess. But having a roaring ocean in an empty shell was even more mysterious.

"I wonder," Newberry thought, "if the ocean roars only in empty shells."

While he was puzzling over this new mystery, a hermit crab stopped nearby to eat some small wiggly things on a pebble. Newberry remembered that hermit crabs had no shells of their own but lived in empty snail and whelk shells. He had once seen two hermit crabs fighting over an empty periwinkle shell. Those two were the homeliest and scrawniest things he had ever seen. Their front end looked like crabs all right, but their back ends looked like curlicue worms. Both crabs tried to back into the shell at the same time. When that did not work, they both came out and had a terrible fight—just like some small boys that

Newberry had seen on the beach. Finally the larger crab managed to slip into the shell before the smaller one. The loser, very embarrassed at having nothing on, hid behind a pebble to make sure that nobody was looking before he sneaked away in search of a shell for himself.

"Hello!" Newberry called to the crab who was having lunch.

The crab kept right on eating.

"Hello!" he called again.

"Hello, yourself," mumbled the crab between mouthfuls. What a messy eater he was! Bits of food were dropping all around him.

"Can you tell me if the ocean is roaring in your shell?" Newberry asked.

At this funny question the crab stopped with a piece of food halfway to his mouth. "There is no ocean in my shell, just me."

"What I mean is, can you hear any roaring in your shell?"

The crab looked around, first at Newberry, then at the flat, calm ocean at the edge of the mudflat. "The ocean is as calm as a puddle."

"I know," replied Newberry, "but just the same would you mind ducking back into your shell and telling me if you hear anything?"

The crab thought that Newberry had surely lost his mind, but just out of curiosity he ducked as far back into his shell as he could. As he expected, all was quiet—so quiet that he promptly fell asleep.

"Do you hear anything?" called Newberry after waiting a while. There was no answer so Newberry squirted a small fountain of water at the crab to get his attention. A sleepy answer with lots of echoes came mumbling out of the shell, "No, no, no, no, no."

This was most disappointing. The crab had not been very helpful. Then Newberry had a grand idea. Why not listen inside his *own* shells? He pulled in his neck and closed his shells as tightly as he could. This did not quite work. He had some difficulty because lately he had begun to grow fat from too many between-meal snacks.

" I must do something about that," he thought. " I need more exercise." It is not easy, however, to get much exercise while lying in the mud doing nothing day and night. Besides, Newberry was lazy.

He tried again to close his shells. When he clamped the front end of himself shut, the rear end bulged out. When he tried to tuck in his rear end, the front end popped out again. So he squirted out some water to make himself thinner. That helped a little bit, but getting completely in was still quite a trick. At last he succeeded in tucking himself in on all sides. Then he listened.

Everything was quiet. It was so quiet that, like the hermit

crab, he fell asleep even though he was terribly cramped. After a nap full of dreams about roaring oceans, he awoke. It took him a few minutes to remember why he was tucked in this time of morning, but at last it all came back to him. He was listening for the roar of the ocean.

He listened until his head ached. At first there was not a sound. Then he began to hear something like surf pounding on a beach: *kerthump! kerthump! kerthump!* He grew terribly excited. As he got more and more excited, the sound went *kerthump, kerthump, kerthump, kerthump*—faster and faster. Finally Newberry realized that he was hearing his own heart beating. As soon as he discovered this, the kerthump slowed down.

Newberry sighed. This was all very disappointing. He sighed again. As he did so he heard very faintly a different oceany sound, something like waves chuckling and gurgling among the rocks just as they often did around the rock near which he lived. This new sound was so real that Newberry pushed out his neck to see if the tide had come in while he had been napping. But no, he had not been napping that long. The tide was still very low. It was a full moon tide; so the water was even farther away from his rock than usual.

Well, he would just have to listen again. So he stuffed himself back into his shells a second time, and, as before, he had trouble tucking himself in all around. He managed finally after giving himself three painful pinches. "I really must exercise more," he reminded himself.

No sooner was he settled in than the sound commenced again—*chuckle, gurgle, bubble, rumble, swish!* This was not exactly a roar, but it had a watery sound like the ocean. In his new excitement Newberry burped. Immediately the sound stopped.

A disappointed Newberry realized that he had been listening to his own stomach gurgling because of his cramped position after his big between-meal snack. There were no other sounds at all—no waves, no ocean, no roaring, nothing.

The ocean did not roar in his shell; it did not roar in the hermit crab's shell. Newberry decided that it could not roar in moon shells either. If people heard the ocean roaring when they held shells to their ears, the sound must be in their own heads. Newberry had a funny thought: would he be able to hear the ocean roar if he listened to a person's head?

Hours later he was still wondering. He never did solve the puzzle. The full tide came up the beach, a fierce gale blew in from the southwest, and as Newberry lay snug in his burrow he could hear from far off the true roar of the ocean as all night long the surf pounded the beach of his island.

9

Lights in the Sea

Newberry had not been sleeping well lately. Each morning he woke up so tired that he could hardly squirt at all. He used to be the champion squirter in the mudflat. When he was in good form he could squirt all the way from his burrow to the nearest big boulder. At low tide it used to be a lot of fun to try to knock the sand fleas off the rock with a powerful squirt. He and the clam who lived next door used to have squirting contests and Newberry always won. But these days he was just too tired. He would squeeze himself at the squirting end until his other end almost popped. The best he could manage was a little fountain that sprinkled back on top of him. The news spread around the mudflat that Newberry couldn't squirt. He was terribly embarrassed. A clam without a squirt was like a bird without a song or a mouse without a tail.

The trouble was that Newberry was having nightmares. In one bad dream he thought he was drowning in a clam chowder. This was ridiculous because Newberry had never even seen a clam chowder. In another dream he popped out of the chowder and landed in the frying pan where he almost became a fried clam before he awoke. Another night he dreamed that he had lost his shells. That was very disturbing.

Newberry's nightmares began when he started eating big snacks before going to bed. A little snack was fine, but then he began to make a regular meal of it. That is when his dreaming started. The worst dream was when he woke up in the middle of the night in his dark burrow. Newberry actually began to be afraid of the dark. His burrow always used to be a snug, safe place to be. Now he began to imagine that in the mud walls of his burrow there were all sorts of monsters and dragons and two-headed clam worms ready to dig through the mud walls and eat him. "If I only had a light in here," thought Newberry, "I would feel so much better."

Later that week the low tides came at night, and since Newberry couldn't sleep he was listening to the crickets and counting shooting stars. Usually the songs of crickets lulled him to sleep and when that did not work counting shooting stars always did. But they did not work tonight. Newberry had already counted thirteen shooting stars and was still too wide awake and too afraid of his dark burrow to go back to sleep.

"Fourteen," he counted, and then a few minutes later, "fifteen," then "sixteen." But star number sixteen did a very unusual thing for a star. It not only fell, it went up again. Then it did a loop and then a spiral. "That's no star," said Newberry.

Then the star came shooting down once more and before Newberry could duck it crashed right on the tip of his neck.

"Whoosh!" said a small voice.

"Whoosh, yourself," said Newberry when he had recovered from his surprise. "You're not a shooting star."

"Don't be ridiculous," said the small voice, "Of course I'm not."

"Who are you," asked Newberry. "I can't see in that bright light."

"I am a firefly."

"What are you doing on my neck?"

"Oh, I didn't mean to land on your neck, but I lost control trying to do an outside loop. Thank goodness you have such a flabby neck."

If Newberry had not been so tired he would have lost his temper; however, these days everybody was making fun of him so he just sighed. When he sighed some gurgly air rushed out of his neck, and the firefly almost fell off.

"What's this ragged thing around your neck?" asked the firefly.

"It's my muffler." replied Newberry proudly.

"Awful color," said the firefly.

"It's your greenish light," Newberry said.

"Don't make fun of my light," said the firefly crossly.

"Don't make fun of my muffler," said Newberry just as crossly.

They might have had an argument, but Newberry was too tired to argue and the firefly still had the wind knocked out of him. Secretly Newberry admired the firefly's light. And secretly the firefly admired Newberry's muffler.

Then Newberry had a wonderful idea. If only he could get the firefly to live with him he would not have to go to sleep in the dark. He mentioned this to the firefly.

"That's impossible," said the firefly.

"Why?"

"For one thing I would drown in your burrow when the tide came in."

"Drown?" said Newberry. "What is 'drown'?"

"I'd get full of water and would not be able to breathe."

Newberry was puzzled. "I'm always full of water and I have no trouble breathing."

"I don't know how you do it. It must be quite a trick."

Newberry thought it was very curious that fireflies could not breathe in the water.

"You know," the firefly said, "There are many creatures in the sea that have lights like mine. Why not invite some of them to share your burrow?"

"I never thought of that. I think I'll start looking right now."

The firefly went back up in the air to practice his outside loops and as soon as the tide came in Newberry began to look for someone with a light to share his burrow. He was astonished at how many sea creatures there were that could flash like fireflies. The first one to come by was a kind of jellyfish.

"Would you like to share a cozy burrow with me?" asked Newberry.

"And get squashed to death!" remarked the jelly fish. "Not me."

The next one to flash by was a wee crablike creature. But he too refused. And so did a couple of worms with bright lights.

Now that he was paying attention to the world around him, Newberry noticed that the whole sea was full of very, very tiny cells that flashed brightly whenever the water was disturbed. If Newberry sucked in a clamful of water and squirted it out, it sparkled with thousands of tiny lights. This gave him a wonderful idea. He filled himself full of water with the tiny cells, crawled deep into his burrow, and squirted the water all around. Immediately his burrow lighted up with sparkling lights just like a Christmas tree. Newberry was delighted. Happily he sucked in and squirted out and his burrow glowed with a cheerful light. Before he knew it, he had dropped off into a deep slumber. As soon as he fell asleep the water in his burrow became quiet and all the little lights went out. But the darkness did not bother Newberry one bit. He was fast asleep and if he had any dreams at all they were happy ones. Never again was he afraid of the dark.

10

The Million-Year-Old Clam

Toward the end of summer there was a terrible northeast storm. It lasted for three whole days. It carried away Cy Jones's lobster car with a hundred lively lobsters in it. It banged Captain Bascom's skiff to splinters on the rocky shore and tore the doctor's wooden sailboat loose from her mooring. She drifted out with the tide and would have gone to sea if she had not grounded on the mudflat where Newberry lived.

All during the storm Newberry had remained snug in his burrow dreaming of the days when he was a baby clam without a shell drifting around in the Gulf of Maine. Clams don't have shells when they are born. As a matter of fact, they are really not born. They hatch from eggs, and they don't look like clams at all.

They are funny looking little stomachs floating with the tide and being gobbled up by everyone else in the Gulf of Maine. The lucky ones finally sink to the bottom and grow a tiny shell.

Newberry dreamt of those days and didn't even come up at low tide to see how the storm was going. There were too many things flying about in the gale, and besides, Newberry was afraid that the wind might blow away his purple muffler.

At the end of three days, the waves became mere ripples. The breeze was so gentle it could not even blow away fluff on dandelions—not that there was any left to blow away. So Newberry decided to stick out his neck. When he tried this, however, he discovered that his burrow was chock-a-block full of sand washed in by the storm. He was greatly annoyed. Newberry got annoyed easily when anything interferred with his lazy way of living. You would think that a lazy fellow like Newberry would be more patient, but he wasn't.

Now he had no choice but to dig himself out. Everything went quite well until toward the end, when he discovered a rather large pebble stuck exactly in the middle of the doorway. Try as he might, he couldn't budge it. Not until two tides later did he succeed in removing the pebble.

Most clams would have lost interest once their burrow was cleared, but Newberry was a curious clam. That does not mean that he was odd or queer; he just had more than his share of curiosity, and it often got him into trouble, like the time he stuck his neck in the old medicine bottle. He always wanted to know more about things that he did not understand.

So instead of paying no more attention to the pebble and just waiting for the next tide to wash it away, he looked at it very carefully. He wanted to know what kind of pebble it was, where it came from, and how it got to the mudflat. Newberry had run across a lot of pebbles in his lifetime, and many pebbles had been rolled across him by the tides. Of course he knew nothing at all about pebbles, but he did appreciate different shapes and colors and sparkles.

This pebble was just a plain disappointing gray. It was not very interesting, so Newberry tried to turn it over by squirting water at it. That didn't work, so he tried squirting water under it. That worked. It worked too well: the pebble toppled over onto Newberry's neck and made a big bruise. He would have lost his temper at this except that he got a great surprise when the other side of the pebble came into view. His surprise was so great that he choked on a squirtful of sandy water. There in the solid pebble was a little clam. It wasn't really a clam, but it certainly was a perfect picture of a clam. Newberry had never seen anything like it before. He didn't know what to make of it, but he soon had a chance to find out.

When the doctor rowed over to the island to get his sailboat, he thought he might as well pay a friendly visit to

Newberry since he hadn't seen him for quite a while. From time to time he worried about Newberry because a lot of commercial clam diggers were around this summer. A bushel of clams was worth forty dollars. Sometimes the doctor had bad dreams about finding a purple muffler in his chowder.

It was no trouble finding Newberry because all the doctor had to do was look across the mudflat until he saw a purple muffler. When he arrived, Newberry showed him the pebble with the little clam. The doctor knew all about kidney stones, gall stones, and the stones that babies swallowed or stuffed up their noses and in their ears. He didn't know much about stones outside of people, but he promised that he would write to the University of Maine and find out for Newberry what the pebble was.

A week passed before the doctor sailed to the island again. He anchored his sailboat in deep water and came ashore in his dinghy, being very careful not to drag it up on top of Newberry.

"Your pebble," he told him, "is a fossil."

"A fossil?" Newberry had never heard of a fossil. "What is a fossil?"

"A fossil," explained the doctor, "is what is left of a clam or anybody who died millions of years ago and left a print of his body, like a footprint, in some mud that later turned to stone."

All of this was too much for Newberry's small brain to understand at once. You mean that clams were living here millions of years ago?" he finally exclaimed.

"Well, maybe not here," the doctor said. "This particular fossil came from up in Canada."

"Where's Canada?" Newberry asked. "Is it near Blue Hill?"

"Oh, no! It's more than a hundred miles away."

Newberry was not very good at geography, and he had

never been a hundred miles anywhere, not even when he was drifting around as a baby clam. He changed the subject.

"How did it get here?"

"The glacier brought it."

"The what?"

"The glacier. The big sheet of ice that came down here from the North Pole millions of years ago."

Newberry's small brain was in a whirl. Some things were just too complicated for clams, even smart clams like Newberry. But one thought stuck in his mind, the thought that clams were millions of years old.

"Are there any fossil people?" he asked.

"People don't live in the mud," the doctor reminded him.

Newberry couldn't understand why anyone would want to live anywhere else.

"How long have people been here?"

"Oh, people came here a few thousand years ago," the doctor replied.

Newberry thought about this for a moment.

"So," he said at last, "Maine really belongs to clams."

The doctor tried to explain that it wasn't quite as simple as that, but Newberry wasn't listening. He was dreaming about a Maine with no people, where there were no such things as chowders, and clams lived happily forever.

11

A Peck of Trouble

Newberry woke up feeling very very grumpy. Nothing pleased him. The water was too warm, the sun too bright, the waves too big, his burrow too tight, and the mudflat too crowded. Grains of sand got stuck under his muffler, his foot was cramped, his neck ached, and his stomach rumbled. There really was no reason for him to be so cranky. He knew he was out of sorts, and he didn't know why. That made him crankier than ever. When his neighbor, the old barnacle, opened his shell and waved his feet and said "Good morning," Newberry just grumbled at him.

The trouble was that life was too easygoing. Newberry had everything to make a clam happy. Most of his troubles were imaginary. He just didn't know what to do with himself. That made him careless. He was so bored that he didn't pay much attention to what was going on around him. He should have been warned by the sight of a hermit crab scuttling away so fast that he almost ran out of the snail shell house that he carried around with him. He should have seen the shadow that suddenly blotted out the sun. The old barnacle tried to warn him, but Newberry was too wrapped up in his own misery to listen.

"Well" the old barnacle said to himself, "I hate to see him go, but it serves him right." So he closed his shell with a snap to be out of harm's way.

The shadow that blotted out the sun was shaped like a man. It was a thick humpbacked shadow, but that was only because the man who made it was hunched over raking clams. He was a commercial clam digger who had come to the island because most of the good flats had already been dug out, except for the baby clams. He had started digging just above the half tide mark.

As the tide ebbed he dug farther and farther down the beach, leaving behind a long furrow full of broken shells, clam worms cut in half, smashed sea urchins, rocks, shrimps, a couple of baby eels, small crabs, a polished piece of green glass, some seaweed, a bit of sea lettuce, part of an old enameled pie plate, a bronze nail from a boat wrecked long ago, a few fish bones, three pieces of ribbon worm that only a moment ago had been one whole worm, and all sorts of other interesting objects.

While Newberry was busy feeling sorry for himself and mumbling and grumbling, he suddenly found himself yanked out of his burrow, picked up from the mud, and tossed into a hod with a peck of other clams. He landed on something hard with such a jolt that a little fountain of water squirted out of him.

Before he had any time to collect his thoughts the hard

thing on which he had landed said, "Can't you be more careful who you land on?"

Still not realizing what had happened to him, and still feeling cranky, he snapped back at the hard thing—which happened to be another clam—"Be careful yourself! Why did you get in the way?"

"I wasn't in the way." said the other clam. "I was here first."

"Well, if you don't like it, move over," said Newberry impolitely.

"Move over yourself."

"I won't. It's my burrow," said Newberry forgetting for the moment where he was.

"Stupid! You're not in a burrow. You're in a clam hod."

All of a sudden Newberry came to his senses. He realized that what he had feared had finally happened. All his grumpiness left him. Even though he was in a hod full of clams he felt very much alone.

"I suspect," said the other clam, "that we will end up in a chowder or fried in bread crumbs."

At this horrible thought, Newberry shuddered so hard that his two shells rattled like a pebble in an empty moon shell.

In the meantime a dozen more clams landed on top of him. It seemed as though every clam in the mudflat was being dug that day. Newberry lost all hope. His luck had run out, and this was surely the end. He thought of all his narrow escapes other times. He remembered how impolite he had been to his friend the old barnacle. If he had had tears for crying, he would have cried.

The next several hours were a nightmare full of frightening sounds, dizzy movements, unfamiliar smells. When the clam digger finally filled the hod, he sloshed it up and down in the

M. 1981

water to wash off the mud. Two clams fell off the top of the pile and escaped. Poor Newberry was down in the middle of the heap. He was certain that this was the last time he would ever feel the familiar Gulf of Maine water.

When the hod was dumped with a thud into the skiff, all the clams were banged about so that as they tried to close their shells they pinched each others necks.

"Let go of my neck," cried Newberry.

" I will when you let go of mine," replied a big gray clam.

Next, the boat began to vibrate, and a terrible tingling ran through all the clams. Never before had Newberry experienced such an itchy, tingling feeling. Since he didn't have arms and legs, he had no way of scratching. On top of it all, the outboard motor that was causing the vibration gave off a sickening smell of oil. Because the skiff was headed toward Blue Hill and traveling with the wind, the fumes were blown full strength into the boat. Newberry started to lose his dinner but then remembered that he had not eaten any dinner. His insides were empty. Just the same, he felt as though he were going to turn inside out.

When the boat finally landed in the inner harbor, things began to happen very rapidly. The clam digger carried the hod to the fish store on the town wharf, where he sold the whole lot to the storekeeper. The storekeeper dumped the clams on a bed of cracked ice. Then he stuck a sign in the ice. It read, STEAMERS.

For the rest of that day and most of the next, Newberry was more humiliated than he had ever been. Mrs. Thrumble came in and asked how much the steamers were just as if they were things like cucumbers and cabbages.

"Four dollars a peck."

"That's too much," she said, "give me a pound and a half of scrod."

Old Mrs. Peabody poked at Newberry in an unfriendly

way as she asked, "Are they fresh? The last time I got some bad ones."

"Just dug yesterday," said the fisherman.

"This one looks dead."

"Can't be. Just came out of the bay."

"I hope they're not polluted. Everything's polluted these days. Don't know what the world's coming to."

"Nope, they wasn't dug in the harbor."

But then Mrs. Peabody bought a lobster instead.

Later in the day, two small boys poked at the clams while their mother was busy looking at haddock.

"No, not that piece," she said. "There's a fly on that one."

One boy picked up a big clam lying on top of Newberry and squeezed it so that it squirted at the other boy. The other boy grabbed Newberry and squeezed him. Newberry tried his best not to squirt, but the boy squeezed so hard that Newberry couldn't hold back any longer. Newberry got his revenge, though. Instead of squirting from the end that the boy was pointing toward his friend, Newberry squirted from the other end and got the boy who was teasing him right in the eye.

Just before the store was ready to close at the end of the day, one more customer came in and asked for a peck of steamers. Newberry recognized the voice. It was that of his friend the doctor. As the fishman began to dump clams into a brown paper bag, the doctor spied Newberry's purple muffler.

"Newberry!" he cried. "What on earth are you doing here?"

Newberry hung his neck in shame.

"I'm being sold," he said.

The doctor was very embarrassed.

"Don't worry," he said. "If you don't mind sleeping on a bed of seaweed in my root cellar tonight. I'll row you out to the island before I go to the hospital tomorrow morning."

The next morning the doctor was as good as his word. By the time the tide came in, Newberry was back in his beloved mudflat, snug in a new and better burrow under the same rock where his friend the old barnacle lived. It was a long time before Newberry was ever grumpy again.

12

What Happened to the Mudflat?

One bright morning the Coastal Fisheries Warden came to Newberry's island in his Boston Whaler. He ran the boat ashore at the edge of the mudflat not two yards from where Newberry lived. After throwing his Danforth anchor into the mud, he waded ashore. He walked up to a large paper birch that leaned out over the rocky ledge where the mudflat became dry land. He carried a hammer, some roofing nails, and a piece of cardboard with some printing on it. It only took him a minute to nail the cardboard to the tree.

When he had finished, he waded back to the Whaler, hauled the anchor aboard after carefully washing off all of the mud, and putt-putted around to the other side of the island to another, smaller mudflat.

Newberry was dying to know what the new sign said. He stretched out his neck to discuss the matter with a young clam who lived in the next hole. He got quite a shock when his neighbor stuck his neck up.

"You look positively green," said Newberry.

"You don't look so pretty either," replied the other clam. "You're quite green yourself, and it is not a shade that matches that shocking purple muffler you insist on wearing."

"How do you feel?" asked Newberry ignoring the remark about his muffler.

"Not too good, really," admitted his neighbor. "To tell the truth, I feel horrible."

"My brain aches a little bit," said Newberry. "I wonder what is wrong."

"Just yesterday I was careless enough to be out when a clam worm came by and you know he gave me one look and turned away. He didn't look too healthy himself."

"Well," said Newberry, "When you get a chance, ask around and find out how the other chaps in the mudflat are doing."

Two days passed and Newberry's brain ache did not improve one little bit. Not only that, the news was bad everywhere. Whatever was the matter with him seemed to be the matter with everbody who lived in the mudflat. The clams were turning green, the shrimps had become as red as if they had been cooked, the worms faded to an unhealthy white, the moonshells got so soggy that they could not hold their shells up and kept toppling over. Sea urchins began to lose their spines. One or two became quite bald. One poor starfish had the tips of all five legs curled up so far he could hardly crawl.

Before long, the news got around that there was something very wrong with the mudflat. Nothing like it had ever happened before. The oldest inhabitant, a ragged old sponge

m. 1981

who had been beaten by the waves for many years, nibbled on by fish, and chewed by crabs, said quite positively that the mudflat had never been this way before.

Everybody who could leave began to go. They didn't know where they were going, but any place was better than where they were. Even a sulky old sea anemone who had never been away from the rock on which he sat and never had anything to say, finally cast off and went away in a series of somersaults—to the amazement of Newberry who never knew that the old fellow could move at all.

With each outgoing tide more of the inhabitants departed. There was a steady stream of crawling, swimming, somersaulting, floating animals, until at last nobody remained except the barnacles and the mussels, who were the least venturesome lot, and most of the clams. Newberry was tempted to go, too, but home is home, and who knew what dangers lay in wait beyond the mudflat.

In one way life became a bit safer because with everybody leaving, the crabs and clam worms gone, the sandpipers and seagulls no longer found the mudflat worth visiting. If only Newberry did not have an aching brain, he might have enjoyed the more quiet life. Even the clam diggers stayed away. They stopped coming when they saw the Warden's sign. Newberry felt sure that the sign had something to do with all the trouble. He wished he knew what it said.

Then one day a fish crow landed near Newberry and looked haughtily around. Usually Newberry did not have much to do with fish crows. It was not very healthy staying out to talk to him, but Newberry's curiosity got the best of him. He squirted a weak little stream of water at the crow to attract his attention. The crow cocked his head in Newberry's direction.

"What are you squirting at me for? That's a good way to get yourself eaten."

"I had to take that chance," said Newberry.

"You're a pretty sad looking specimen of a clam. This is the first time I have ever seen a green clam. And with that purple muffler you look ridiculous. The colors clash. You have absolutely no taste in colors. Anyway, you're safe. I would probably get sick if I ate you."

"Please," said Newberry, "have you seen that new sign on the birch tree over there?"

"It isn't new," said the crow in a gravelly voice. "The Warden nailed it up two weeks ago."

"So you did see it."

"Of course! Do you think I'm stupid? My life wouldn't be worth an empty clam shell if I didn't notice everything that goes on around here."

"Please," said Newberry again, "will you tell me what it says?"

The fish crow was immensely pleased that Newberry thought he was smart enough to read. "I'll take a look," he replied.

He flew over to the sign. First he peered at it from the right, then from the left, then right side up and upside-down. When he returned to the mud where Newberry was waiting impatiently, he said, "The printing is rather blurred from last night's rain so I can't make it out exactly." He was making excuses so Newberry wouldn't know that he did not know how to read. "There is a big word I've seen on many other signs, and wherever there is a sign with that word, people no longer do whatever it was they used to do. So that word means 'prohibited.' It means that whatever people used to do here on the mudflat they can't do any more."

The crow knew perfectly well what people did, but he wanted to hear Newberry say it. He loved to tease.

"They dig clams," said Newberry.

CLAM DIGGING
PROHIBITED!
M. 1981

The crow chuckled. "You're safe now. I think your jolly old mudflat is polluted."

"That must be it," thought Newberry. "that is why my brain aches and why everybody who could leave the mudflat did leave." Out loud he said, "I wonder how it came about."

"I can only guess," replied the fish crow, "but I bet you a cob full of corn that it is all because of the new copper mine that opened last year. Whatever it is, it has left this part of the bay very empty. I shall not take the trouble to hunt over this way any more. If you had any sense you would leave too."

But home is home. Newberry stayed on stubbornly. Then one morning late in November, Newberry woke up feeling wonderful. His brain ache was gone. He didn't look green anymore. Better still, the water of the incoming tide felt wonderful and tasted wonderful. A copy of the local newspaper, *The Packet,* floated by with the print side down. Newberry couldn't read the headlines that said that the mine was all mined out and had closed. But he didn't have to—he just knew that the mudflat was all right again. He wrapped his old worn muffler snugly around his neck and settled down for a long, cozy winter.